Yury Avdeev

Green spaces in the urban environment

Yury Avdeev

Green spaces in the urban environment

ScienciaScripts

Imprint
Any brand names and product names mentioned in this book are subject to trademark, brand or patent protection and are trademarks or registered trademarks of their respective holders. The use of brand names, product names, common names, trade names, product descriptions etc. even without a particular marking in this work is in no way to be construed to mean that such names may be regarded as unrestricted in respect of trademark and brand protection legislation and could thus be used by anyone.

Cover image: www.ingimage.com

This book is a translation from the original published under ISBN 978-3-659-89963-8.

Publisher:
Sciencia Scripts
is a trademark of
International Book Market Service Ltd., member of OmniScriptum Publishing Group
17 Meldrum Street, Beau Bassin 71504, Mauritius
Printed at: see last page
ISBN: 978-620-3-52974-6

Table of Content:

Green spaces in the urban environment (the example of Vologda)

Introduction

One of the most important problems of modern landscape construction is to create a system of green spaces in cities and towns. These plantings help to optimize the sanitary and hygienic living conditions of the population, enrich the architectural and artistic appearance of settlements.

With the growth of urbanization, intensive industrial production and motor transport, the concentration of toxic gases, dust, smoke and soot in the atmosphere of cities is increasing extremely rapidly. To a greater extent, the function of protection from the adverse factors of the urban environment is assigned to greenery. They can significantly clean the air from dust, soot, soot. Small solid particles are deposited on the leaves of trees, shrubs and lawns.

Green plants greatly reduce the harmful concentration of gases in the air.

The air temperature among green spaces, especially in hot weather, is much lower than in open spaces.

Green plants, protecting the soil and wall surfaces of buildings from direct sunlight, protect them from severe overheating and thus from rising air temperatures.

Green spaces also have a great influence on improving the radiation regime in the city. [12].

The degree of mitigation of the radiation regime in landscaped areas compared to open spaces is affected by the size of the landscaped area, as well as the density of tree and shrub plantings.

Green spaces can significantly reduce the city's noise pollution. The crowns of deciduous trees absorb 26% of the sound energy falling on them. [5].

However, improper placement of green spaces in relation to sound sources can have the opposite effect, i.e. increase noise levels where they need to be reduced. This can happen when planting trees with dense crowns along the axis of the street with heavy traffic. In this case, green spaces will play the role of a screen, reflecting sound waves in the direction of residential buildings and recreation and sports areas [16].

This monograph examines the issues of green building, analyzes the current state and issues of inventory of green spaces in Vologda.

Natural conditions of the city of Vologda

Climate, soil and hydrography have a special influence on the growth of certain types of trees and shrubs in the city of Vologda.

The Vologda region, including the city of Vologda, has a climate peculiar to the southern taiga, i.e. moderately warm summers and cold winters. The average annual temperature in Vologda is +2.4°.

Peculiarities of atmospheric circulation peculiar to the northern half of the Russian plain have a great influence on the climate of the area. By definition of L.S. Berg, the general circulation of the atmosphere is understood as other air currents, through which the exchange of air masses between low and high latitudes takes place. Any change of air masses usually entails a change in weather and, ultimately, climate.

The transport of continental air of temperate latitudes prevails over most of the territory of the Russian Federation. In summer, as a rule, it is rather strongly heated and characterized by relatively high temperatures, moderate dryness and dustiness. In winter, with clear skies, it cools down considerably over the snowy expanses.

In contrast to the continental air of temperate latitudes, temperate sea air flows to us from the Atlantic Ocean, which is characterized by high humidity and moderate temperatures.

Finally, it is necessary to note the subtropical air that usually comes in summer (though quite rarely), which comes from the southeast of the European part of the Russian Federation, from Kazakhstan and Central Asia. This air is strongly heated, very dry and contains a significant amount of dust.

The weather, and consequently the climate of our area, is influenced by all of the above-mentioned air masses. This results in unstable, variable weather characteristic of the area.

However, continental and maritime air masses of temperate latitudes predominate. If in winter, the area is under the influence of continental air of temperate latitudes, which comes in the form of southwestern flow from the "Voyeikov axis" (Voyeikov axis is a high pressure band formed in winter as the western spur of the Siberian anticyclone and passing within the European part of Russia, approximately along 50° north latitude. Cold and heavy air from the Voyeikov axis flows north and south, deflecting as it moves to the right), then rather cold, frosty weather is established. The air temperature may drop to -15, -20°C in such cases.

Due to the northern location of the area, invasions of cold arctic continental air masses from the north and northeast are also quite frequent in winter. In such cases air temperatures may fall to -30, -35° and unstable weather is replaced by clear, frosty weather. In general, winter in the Vologda area is cold and prolonged, but with thaws, which is especially characteristic of recent years.

In summer, the influence of Atlantic cyclones weakens, and continental air of temperate latitudes, formed locally or brought in from neighboring areas, becomes decisive. Warm but not hot weather with cumulonimbus clouds and local rainfall is common in our summer, unless foreign air masses arrive. In the case of cyclonic activity from the Atlantic, which also occurs in summer (although on a smaller scale), it becomes colder, cloudy and sometimes rainy. In some years, summer cyclones often pass over the area, creating long periods of inclement weather [1].

The Vologda region, including the city of Vologda, is characterized by a moderately continental climate of the forest zone, with moderately warm summers, long moderately cold winters and unstable weather patterns.

Occupying a high-latitude position within the temperate belt, the region receives on average a relatively large amount of heat from the sun per year: 74 b.cal. per 1 sq. cm. of the surface. In comparison with the southern city of Tashkent - up to 102.7 b.cal., a relatively small difference in the amount of heat. Duration of the day in the summer period reaches 19 hours, the maximum height of the sun at this is 54°10' min. and in the winter period duration is 6 hours. Weather regimes are unstable, due to the variety of air masses penetrating to us, their transformation and frequent passage of cyclones, especially in the autumn-winter period.

The climate is characterized by a relatively long warm period of the year, reaching 200 days.

The frost-free period is short, 115-125 days. Winds prevail in the southwestern direction, with an average wind speed of -4.1 m/s [2].

Most of the residential area of the city has an artificial soil cover. Only in parks and gardens the soil layer is preserved in a relatively unchanged form and is represented by sod-podzolic soils.

The modern soil cover of the city is quite diverse and is represented by the following types of soils: sod-podzolic, sod, floodplain and bog, which differ from each other morphology, water-air, thermal, agrochemical properties.

The eastern suburbs, located in the Sukhonskaya Depression, are covered with alluvial deposits. Turfy-gley soils are common on the terraces, alluvial-sod soils and boggy soils are common on the floodplain. On the slopes of the Vologda Upland, where the western and southern suburbs of the city are located, sod-podzolic soils on the cover loam, formed from highly podzolic forest soils, prevail.

In most of the city soils are slightly acidic, close to neutral slightly alkaline reaction.

On both banks of the Vologda River, in its lower reaches, grassy bogs with a thick (up to 1 m) layer of peat and peat-moam-gley soils with peat thickness up to 50-70 cm prevail [3].

The hydrographic network within the city and its surroundings is represented by the Vologda River, which divides the city into two almost equal parts, and its tributaries: the Toshnya, Shogrash, Sodima, Evkovka rivers and several brooks. The density of the hydrographic network is 0.8 km/km 2.

The groundwater level is subject to sharp fluctuations associated with fluctuations in the water level of the river. In different districts of Vologda, the groundwater level varies from 0.5 to 4.50 meters.

All rivers have well-defined river valleys with steep slopes ranging from 5 to 15 degrees.

The width of the river valleys is significant, indicating the duration of their formation. For example, the largest tributary of the Vologda - the Toshnya River near the village of Novoye has a valley width of over 1 km with a channel width of 20-25 m.

The rivers of the district, as well as all the rivers of the region, are fed by mixed: they receive water from spring snowmelt and rainfall. Some role is also played by groundwater feeding, especially in winter, when rivers are frozen and there is no surface flow [2].

Retrospective analysis of green spaces in Vologda

In the old days, Vologda was a city of forests. As the traveler approached, he was struck by the abundance of temples and greenery.

In 1791 the master plan of the city was approved, in accordance with which in February 1822 it was decided to lay out the boulevards along Dvoryanskaya (Oktyabrskaya street), Gostinodvorskaya (Mira street) and Peterburgskaya (Leningradskaya street) streets. The construction of these boulevards required large sum of money for that time - 5520 rubles. The city was not able to allocate this sum, so the City Council organized subscription for voluntary donations for this construction.

The history of Vologda's garden art monuments is of great interest, since in some gardens of the city as early as 1860 there were buildings (arbors, rotundas), which later disappeared. For example, on the Sobornaya Embankment (S. Orlov St.) of the Vologda River. In 1823 by order of the Vologda civil governor N.P.Brusilov a garden was laid out for the donated money. In 1860 there were two arbors, semicircular and quadrangular, built by individuals. Lime trees, poplars, pines were planted, but in our time, due to archaeological excavations, many trees died, on the Sobornaya Hill had to renew the plantings.

One of the oldest parks in Vologda can be considered the Bishop's apple orchard, which was planted before 1688, but in 1770 was partially cut down because the apple trees did not take root and did not bear fruit. At the end of the XVIII century linden trees were planted on the territory of the Metochion, which even today neighbor the building of Joseph the Golden, the chambers of the Government office, the high bell tower of St. Sophia cathedral, perfectly complementing the unique architectural ensemble.

In 1824, in time for the arrival of Emperor Alexander I, the Alexander Garden was laid out. This garden was then a place of rest for the privileged public. On Sundays and holidays the military brass band played here and the citizens walked around. In 1853 the garden was surrounded by wooden bars. Today it is Komsomolsky Park on Revolution Square, where the eternal flame burns in honor of the Vologda citizens who died during the Great Patriotic War.

In 1842 by order of Governor Bolgovsky on Gostinodvorskaya square (Mira street), starting from Malaya Blagoveshchenskaya street (Blagoveshchenskaya street) to the bridge on the embankment of the Zolotukha river, a beautiful boulevard was also arranged. At the corner of Peterburgskaya and Dvoryanskaya Streets, behind the boulevards, by order of the governor's office and with the money of the municipal treasury in 1858 a new garden was constructed. In 1860 there were made paths for 950 sazhens, but there were no trees and shrubs yet. In the garden was dug out Pyatnitsky pond that was used as a fire pond and in winter was cleared for the city skating rink (nowadays the pond is filled up and there is a stadium of VGPU in this place). The garden was intended for public festivities,   existed until the beginning of the twentieth century   .

In the XIX century in the city was arranged Samarinskiy garden, which belonged to a large landlord Samarin. It housed a special building of the penal colony for juvenile offenders. In the 20s of our century, it existed as a Railway Garden, and now the remains of this garden are preserved as a square near the Palace of Culture of railway workers.

In 1872, on the 200th anniversary of Peter the Great's birth, the zemstvo bought from the Vologda merchant Vitusheshnikov a house of the merchants Goutmanov, where Peter the Great had stayed in 1724. Around this house the place was levelled and a square was laid out, which was surrounded by an iron lattice. In 1885 a museum was opened in the house. Thanks to the decorative planting of trees, shrubs and a large flower bed, this historic corner of our city has become very beautiful.

In the last century, street landscaping was also carried out in the city - Kalashnaya (Gogol Street), Kozlenskaya (Uritskogo Street) and Arkhangelskaya (Chernyshevsky Street) streets were solid green corridors.

In 1910, a park was laid out in the center of the city, which was later called the Children's Park. Mostly poplars were planted (with cuttings purchased from private individuals), later - linden, elm, and pine. The park became a favorite vacation spot for townspeople.

The post-revolutionary period in Vologda also saw widespread work on landscaping and beautification of the city. For the first time a spring holiday - the "Day of the Forest" - was arranged on May 6, 1925. At the same time, the Archbishop's Garden (the garden of the Second Russian Recycling Plant) was enlarged, plantings were made in the Alexander Garden (later - Mikhailovsky, now - Komsomolsky) and in the Children's Park on

Revolution Square.

In 1920, the Vologda Museum of Local Lore was given the Bishop's farmstead. In 1923, on the territory of the former Bishop's Garden at the museum the Botanical Garden was founded with an area of 4 hectares, the only one in the Northern Territory. At the end of the 1920s, part of it became the VRZ Culture and Recreation Park.

The idea of creating such a garden in Vologda was born in the Vologda Society for the Study of the Northern Territory in 1915, but only in 1924, thanks to the funds released by the State Executive Committee (600 rubles), the Vologda State Museum was able to implement it in full.

Vologda Botanical Garden had a woody section of broad-leaved and small-leaved species, a hill with shrubs and pines, ponds and about a hectare of plantations of various herbaceous and shrub plants. This area was essentially the Botanical Garden, as the remaining 3 hectares were transferred to the temporary use of the workers' organization - the railway club - and no systematic work was carried out there.

In 1927, more than 80 percent of the plant species registered at that time in the Vologda Province, and 20 percent of the species characteristic of the Northern Territory as a whole, grew in the Botanical Garden. There were 323 plant species in the herbarium section, medicinal - 39, forage - 32, technical and vegetable - 52, trees, shrubs, semi-shrubs and ornamental grasses - 78, weed species - 32. In total, there were over 580 plant species in the Botanical Garden. Plots of natural phytocenosis were created. In the alder grove in 1927, willow, alder, rowanberry, bird cherry, buckthorn, other bushes and shrubs, 30 species in total, were planted. There were also mounds of berries (blueberries, blueberries, lingonberries), herbaceous vegetation, and mosses. During the whole vegetation period, herbaceous plants characteristic of alder forests were planted to form a characteristic cover in them in the future. Similar work was carried out in the "spruce forest.

Thus, two natural groupings of vegetation of the Northern Territory were formed in the Botanical Garden. Seeds of 35 plant species from Samara province were sown in the acclimatization plot, of which 14 species bloomed and bore fruit. Unfortunately, the Botanical Garden in Vologda existed only until 1929.

A detailed study of Vologda's tree plantations was first carried out by local historian A.P. Belizin in 1927. As of 1927, there were only five public gardens in Vologda (railway garden, VPRZ garden, Alexandrovsky garden, Children's garden, and Peter the Great garden, which was located opposite Petrovsky Dome, on the opposite bank of the river). They were relatively small and distant from each other. The number of city boulevards was also limited: there were only three of them - Myasnoryadsky (Mira street), Bolshoy (Oktyabrskaya and Leningradskaya streets) and Soborny (Sobornaya hill). In all gardens and boulevards, there were 17 species of trees, of which 5 were conifers. Only 5 species (birch, linden, sweet-scented poplar, smooth and mountain elms) were widespread, while 12 species were rare or isolated and could not considerably influence the diversity of tree plantings in the city's gardens and boulevards.

Individual species were represented differently in plantations: birch 66.3 percent, smooth elm and mountain elm 8.7 percent, small-leaved linden 4.8 percent, sweet-scented poplar 16.6 percent, other species were insignificantly represented (summer oak, silver poplar, common ash, well adapted to the city conditions, occupied in the percentage ratio one of the last places).

The birch is still the main species used in landscaping in Vologda.

By the project and under the direction of engineer V.I. Sokolov in the center of the city on the Soviet square in 1934-1936 the garden named after S. M. Kirov was laid out. The initiator of its founding was gardener Dorofei Ivanovich Teterin.

In 1936, according to the project of the same engineer, the boulevard was reconstructed from Stalin Street (Mira Street) through Oktyabrskaya and Leningradskaya Streets. The nursery of 3 hectares on Podlesnaya Street (Gorkogo Street) served as the basis for landscaping. The workers themselves collected and dried the seeds for the nursery, as there was almost no money to buy planting material. Elms from the nursery were planted along Oktyabrskaya and Papanintsev streets (Pobedy prospect), birches were planted with large seedlings brought from Pyatnitsky pond, due to the re-equipment of this place for sports grounds. The old birches that fell out were replaced by poplars, which were brought from the village of Kuvshinovo.

In 1936, under the leadership of D.I. Teterina, the square in front of the Batyushkov House and the children's hospital along Victory Avenue from Batyushkov Street to Maltsev Street was also laid out. Smooth elm, linden, and acacia trees were planted. In 1937-1938, plantings were made along Stalin Street and Soviet Avenue.

On May 18, 1939, two thousand citizens of Vologda participated in planting elms along the Vologda River bank, thus laying the foundation of the Central Park of Culture and Rest (Peace Park). The territory of the railway polyclinic was improved with the funds of the Department of Railway Transport, and the garden of linden, elm, apple and acacia trees near the 10th Anniversary of the October Revolution Club (DKZh) was reconstructed on an area of one hectare.

During the Great Patriotic War, the city's plantings suffered to a large extent, as the old trees were cut down for firewood, and the young ones were neglected. It is true that during these years an attempt was made to green the banks of the Zolotukha River, but it was unsuccessful. In the green areas of Zarechnaya and Zavokzalnaya parts of the city were located military units with cars and armored vehicles, which affected the condition of trees: hundred-year-old linden trees along Svyazi Street died.

In the postwar years, work on the improvement and landscaping of the city resumed. Collectives of many enterprises organized subbotniks: to lay the garden on Sobornaya Hill, landscaping of Lenin, Peace, October, Chernyshevsky streets and others. At the same time there was laid the public garden along the Kremlin wall, as well as at the crossing of Gogol and Chernyshevsky streets and behind the theater on October street. Only in 1945, were planted 10 thousand trees, shrubs and ornamental herbs. In 1947 the obelisk of the 800th anniversary of Vologda was opened, and acacia bushes and apple-trees were planted on the Lazy ground.

In autumn 1953, was laid square on Revolution Square, and in 19541955 created the Theater Square on October Street (from the building of the former Drama Theater to Batiushkov Street) and Pionersky - on Soviet Avenue in front of the House of Pioneers (Children's Musical Theater).

In 1954, 87,000 trees, 27,000 shrubs and 20,000 annual and perennial herbaceous plants were planted. In the same year, in accordance with the project of the architect M. P. Amchislavskaya, planting continued in the square on Revolution Square. In 1956, planting began in a new Zarechny park, and from 1955 to 1958 carried out landscaping of the banks of the river Zolotukha. During these years were created green areas in the flax mill, where the workers and employees had planted 28,000 trees and shrubs, planted two public gardens. Collectives of VPRZ, Machine-Tool Plant, Mezhsomolmash, Severny Kommunar Plant and many schools initiated the planting of greenery on their territories.

In 1956, the laying of a park of 4.5 hectares in the Zarechny part of the city began (Veterans of Labor Park (Fig.1.3)); in 1958-1959, squares were laid in the October settlement, Schmidt Square, VI Army Embankment, near the locomotive depot, shipyard, the "Northern Communard" plant.

In 1959, a long-term landscaping plan for 10 years (1960-1970) was made, which provided not only for landscaping of the ancient city, but also improvement of its architectural and landscape plan. According to this plan the planting of the Peace Park was continued, its area was increased, the square near the monument to the 800th anniversary of Vologda was reconstructed: apple-trees and acacia bushes were replaced by cotoneaster, flower beds were laid out. In front of the building of the regional executive committee 17-year-old apple-trees, roses, cotoneaster were planted instead of dry trunk birches; flower beds were created.

In accordance with the landscaping plan of the city, work was carried out on the landscaping and landscaping of the VI Army Embankment from the House of Culture to the Red Bridge: lime trees, spirea, barberry, and poplars were planted. In 1960-1962, the reconstruction of Pushkinskaya Street was completed with the creation of the boulevard. There were planted apple trees, linden trees, shining cornel, blue spruces. At the same time on Predtechenskaya street (Menzhinsky street) - from the Soviet avenue to Herzen street - young oaks were planted. During the same years, the staff of the "Northern communard" plant laid out another public garden on the bank of the Vologda river opposite the central gatehouse, the plant's territory was also planted with greenery.

In 1962, was founded Victory Park in Zavokzalnaya part of the city (near the stadium "Locomotive"), and in 1964-1966. the reconstruction of green areas on Blagoveschenskaya Street (K. Zetkin Street), Samoilo (Siberian Street), Voroshilov (Galkinskaya Street), landscaped area of the regional hospital, pharmacy warehouse, the valley of the Shogrash.

In 1967, the 50th anniversary of October was completed reconstruction of Mira Street from the gym "Trud" to Wintersky bridge connecting Herzen Street to Oktyabrskaya Street (in two rows were planted trees and shrubs), completed landscaping Gorky Street and Leningrad highway (to the exit from the city), along which were planted young spruces.

In the work of V. I. Sokolov, an engineer and architect of the city of Vologda, one can find data on public plantings in the Children's Park, the Park of Labor Veterans, the Peace Park, the recreation garden of Lukyanovo settlement, where the dates of the establishment of these parks and the plantings made there are indicated. According to his data, birch, poplar, spruce, larch, maple, and ash trees grew in the city's public gardens and parks; of shrubs - lilac, caragan, various types of spirea, hawthorn, honeysuckle, rose hips, cotoneaster, and others. Inside the quarters, along with birches and poplars, grew apple trees, cherry trees, rowanberries, and bird cherry, and of shrubs - guelder rose, raspberries, gooseberries, black and red currants.

In 1972 the cornerstone of Vologda's 825th Anniversary Public Garden was laid along Prokatov Street in front of the Vologda 800th Anniversary Bridge. Workers and employees of the "Dormash" plant, students of construction and agricultural colleges, and students of School No. 17 took part in its laying. Thanks to efforts of Vologda Department of Local Railways the Victory Park was continued near "Lokomotiv" stadium, but because of the lack of care only 60-70% of plantings took root. In addition, the species for landscaping this park were not quite well chosen, without taking into account the fact that most of the park's territory is flooded by groundwater in the fall and spring. At the same time, blue spruces and cotoneaster were planted in the square near the monument to aircraft designer S. Ilyushin, and a flower bed was laid out. In addition, the Komsomolsky Garden was significantly reconstructed and blue spruces were planted near the Eternal Flame. Through the efforts of Dorremzelenkhoz, workers and employees of the flax mill, Initsiativnaya Street and Tekstilshchikov Street were landscaped and landscaped up to their connection with the Moscow Highway.

From 1960 to 1980, according to the plan for the development of green spaces in the city, squares were planted along the Kopanka River between Zosimovskaya Street (Kalinina Street) and Galkinskaya Street (Voroshilov Street), green spaces were planted along Petina, Blagoveschenskaya, Chernyshevsky, Klubov Streets, Poshehonskoye Highway. In honor of the 60th anniversary of the October Revolution, an alley of 60 oak trees was planted in the Peace Park, a public garden was laid out on Market Square, plantings were made in Tchaikovsky Square, and a public garden was put into operation on the Kopanka River.

In 1979, there were protective zones near gas stations, a protective green zone along the Sodema River. In 1980, a tree and shrub nursery was put into operation.

Thus, thanks to the efforts of many organizations, Vologda acquired a modern green outfit, the range of plants planted was significantly expanded, and interesting compositional solutions were found. [4]

The current state of green spaces in Vologda

Urban conditions have an impact on the ecological condition of the vegetation.

Plants are differently related to environmental conditions: some are light-loving, while others tolerate significant shading, some are demanding to the soil, while others easily tolerate poor soils, some are frost-resistant, while others do not withstand even mild frosts. All of these factors, of course, affect the formation of dendroflora of the city. However, the leading factor in the selection of species is not the ecological characteristics of plants, but the availability of planting material.

During the construction of the city, it was natural to want to green it quickly, so the choice fell on the birch and various types of poplars - species that grow quickly and are less demanding to the environmental conditions. The 1920s and 1950s can be called the birch and poplar years for our city. These species prevailed both in gardens and in the streets. So, according to E. Ispolatov the birch plantations in the park of WWTP were 69.2%, in Komsomolsky public garden - 52.7%, in the garden of USSR Railway Company - 79%, in the public garden near the cabin of Peter I - 70%. The other species were not so numerous, and their diversity was low. Shrub species were even more scarce.

Street landscaping in general was formed spontaneously: the selection of species was random and depended on the wishes of the residents of nearby houses; plantings were located in a crowded manner, close to the roadway and sidewalks. During the expansion and reconstruction of old streets and the construction of houses, green spaces were ruthlessly destroyed, and new ones were either not planted or took root poorly without proper care. As a result, many streets were unprotected, and citizens now suffer from excessive noise and pollution.

By 1988, Vologda had 6 parks, 4 gardens, and 16 public gardens. The largest park (an area of 150 hectares) is Peace Park.

The largest garden in terms of area was the garden of the Oktyabrsky settlement (5.3 hectares), the garden near the City House of Culture (3.67 hectares) was in a neglected state, the industrial district garden (1.3 hectares) was in a satisfactory state, the garden of the House of Culture was in a good state (1.9 hectares). The Babushkin square was the largest (4.8 hectares, this area included the area of the square named after the 825th anniversary of

Vologda, previously laid out in this place), the second largest - the square on Tchaikovsky Square (3.4 hectares).

At the end of 1990 in Vologda there were 340 hectares of public plantations, or 12 square meters per capita.

Most public gardens have an area of 0.1 to 0.5 hectares, although according to current standards, the area of the public garden must be at least 0.5 hectares [4].

At present, public plantings are distributed very unevenly across the city's planning districts.

Public plantings are the most poorly represented in the Zarechny part: there is only one park (Veterans of Labor), two gardens (near the House of Culture and the garden of the industrial district), six public gardens and two boulevards. The largest area of green spaces is in the central district. There are two parks (Mir park and Oktyabrsky park), 2 gardens (Kremlin and near the House of Culture), 18 public gardens, and 11 boulevards. The South district has 3 parks (the Victory park, the Osanovskaya Grove, the Yevkovka district park), 2 gardens (the Oktyabrsky garden and the district garden), 4 public gardens, and 6 boulevards. Appendix A contains a scheme of landscaping of the Municipality "Vologda City".

The landscaping of the city, as already mentioned, is closely related to its development, the increase in urban land area and the number of residents. According to the building codes, the area of public green spaces in medium-sized cities should be 14 square meters per 1 inhabitant. In Vologda at the end of 1938 total area of green spaces was 22 hectares, 95 000 people lived in the city, i.e. 2.3 square meters per 1 person.

According to the inventory data of 1948, there were already 7, in 1957 - 9,4, in 1975 - 11 square meters of green spaces per inhabitant19. In 1980, the green areas of the city were brought to 538 hectares, which in terms of one inhabitant amounted to 21 square meters, but this is taking into account the green area of the city, the area of landscaping for general use was only a third part.

As of 1990, the total area of all green spaces and plantings within the city limits was already 681.5 hectares, but the public areas were only 185.3 hectares, or 6.1 square meters per resident. A clear violation of the norms of providing city residents with green spaces is clearly visible.

An important aspect in carrying out landscaping is the selection of an assortment of species. The species composition of ancient Vologda gardens, boulevards, and neighborhood plantings is very poor - about 20 species of trees and shrubs: various poplars, lilacs, hawthorn, elm, lime, birch, yellow acacia, occasionally larch, fir, ash, oak. The natural conditions of Vologda allow using a much larger number of tree and shrub species, which would provide a proper architectural and dendrological design of the city [4].

At present Vologda takes the last place by the number of species, forms and varieties of tree species used in landscaping of cities in the North-West. This indicates that our city has all conditions for vegetation to have not only ecological but also aesthetic impact on the residents. In this respect the plant world is unique, we only need to use its possibilities correctly.

Problems of green building in the city of Vologda

Urban greening is the most important component in the overall complex of urban planning and urban economy. Parks, gardens, boulevards, squares largely determine the planning structure of the city, are essential elements of its cultural landscape, contribute to the creation of the best sanitary-hygienic and microclimatic living conditions for the population.

The problem of landscaping in the city is manifested in the reduction of the area of green areas, the unsatisfactory condition of existing green spaces, low motivation of landscapers to a high quality of work performed, dissipation and misuse of budgetary funds, which leads to the lack of efficiency of landscaping work.

To date, the city of Vologda has 494.48 hectares of public greenery. Green areas of the city are not in a very good condition. When creating and replenishing them, as a rule, the cheapest and unimpressive tree species are used. There is also no unified base of planting material.

There are several criteria for selecting plants and designating trees for transplanting:
- age of physiological aging of trees. It depends on both species features of trees and their growing conditions. In natural forests, conifers (spruce, pine, larch) and some deciduous trees (oak, elm, linden) live more than a hundred years, in the park conditions trees also retain viability for a longer time than in plantings on the streets and in residential neighborhoods, where they are exposed to a much greater negative impact of the urban environment. On average, physiological aging for different species of trees in urban green spaces occurs: oak, elm and chestnut trees at the age of 80-90 years, linden, ash and Norway maple at the age of 70-80 years, pine, larch and cedar at the age of 80 years, spruce at 60 years, birch, pear, mountain ash, alder at 60 years, poplar and apple trees at 50 years.
- damage to green spaces caused by various natural and anthropogenic negative factors. Specific diagnostic signs have damage caused by many pests and manifestations of many infectious diseases, as well as consequences of some negative factors of natural and anthropogenic nature (action of frost, drought, violent winds, atmospheric and soil pollution, etc.).

With the development of industry, an increase in the area of residential development, the area of green spaces is decreasing.

In 1994 the State Enterprise "Aerogeodesy" of St. Petersburg performed an aerial survey of the city of Vologda on a scale of 1:5000. The aerial survey resulted in a topographic plan on a scale of 1:2000.

Using the example of two aerial photographs, an inventory of green spaces in the Zavokzalny and fifth microdistricts was conducted.

When examining the aerial photos under a stereoscope, we distinguished massive green areas, as well as freestanding trees.

The total area of the study area was 184 ha. The total area of green spaces was 34 hectares. This is 18% of the total area.

The largest number of green areas is located in the Zavokzalniy Microdistrict, as the area of old residential development. There are also school № 29 and boarding school № 2, which are surrounded by green areas. In this neighborhood flows the river Shogrash, along the banks of which there are many bushes and trees. The fifth neighborhood is relatively new. Therefore, there are fewer green areas than in Zavokzalnoye.

Along the linear objects: Mozhaiskogo Street are planted single-row tree plantings; along the railroad there are also masses of green spaces; along Konev Street there are no tree plantings.

When looking at satellite images, the area of green spaces decreased by 5 hectares in the area of Koneva street. This is due to the fact that the area is rather young and now intensive construction is being carried out there.

In a park like Osanovskaya Grove, the problem is untidy fallen trees and large amounts of trash left behind by vacationers.

Common problems of parks and gardens in the city are:
- neglected green spaces;

- trampled lawns in outlying residential neighborhoods

- lack of flowerbeds in certain neighborhoods of the city;

- the polluted condition of water bodies;

- destruction of the road surface.

Issues of inventory of green spaces

An inventory is a documented record of all garden elements on a given property.

The inventory of green spaces is carried out in order to:

- setting the boundaries of the planted natural area and documenting them;

- obtaining reliable data on the number of green spaces in the city, their condition for urban management at all levels of management, operation and financing, assigning them to the appropriate category of land, protective status and maintenance regime;

- Establishing the species composition of trees and shrubs, determining the number, category and type of plantings, age of plants, diameter (for trees), condition, and areas of lawns and flower beds;

- timely registration of changes that have occurred;

- Determination of land users of territories and establishment of responsible organizations, legal entities and individuals for their preservation and condition;

- Establishing the presence and belonging of stationary engineering and architectural structures and equipment of landscaped/natural areas (fountains, monuments, sculptures, etc.);

- regulating the maintenance of green spaces, their overhaul and reconstruction;

- organization of the rational use of the city's territories;

- to provide electronic accounting of landscaping objects and green spaces in the city as a whole in the maintenance of the Registry of Green Space and green space monitoring.

All green areas, regardless of their organizational and legal form of ownership and departmental affiliation, located within the city limits, with established boundaries and provided for use (possession) shall be subject to inventory,

order)   to the responsible land users   (institutions, organizations, enterprises or individuals).

In accordance with these purposes, the inventory of green spaces is as follows:

J determination of the total area occupied by green spaces;

Establishing the number of trees and shrubs with determination of plantation type, species, age of plants, trunk diameter at 1.3 m height (for trees), and condition;

entering data obtained as a result of the inventory of green spaces into the city's unified information and analytical complex (electronic version);

filling out a record card (passport) of a green area; compiling consolidated data on the green areas of the municipality;

timely registration of changes that have occurred.

The inventory of plantations begins with the collection of ecological information about the site and the numbering of trees. During the inventory, all metric indicators are taken, structural defects, traces of human activity, forest pathology and sanitary condition, recommended works (removal, pruning, treatment, protection, soil aeration) are noted - all for each tree.

Inventory is carried out on the basis of the approved situational plan (scale 1:2000) and topoplan (scale 1:500) in two stages. At the first stage the area, boundaries and classification of the object are established. At the second stage the qualitative and quantitative condition of green spaces and landscaping elements is determined.

Materials are collected during the preparatory phase of the inventory, which includes:

1) setting the goal and objectives of the survey;

2) preparation of materials and equipment for fieldwork (tablets, pencils, rulers, rubber bands, compasses, measuring tapes, tape measures, measuring forks, rope, paper);

3) familiarity with the object of the survey;

4) making a plan-map of the object of the survey (park, square, boulevard, street, etc.) [13].

Landscape inventory is a pre-project landscape-architectural and biotechnical study and assessment of the territory. It is carried out for the purpose of identifying and describing taxation divisions according to their biological, landscape-architectural, sanitary-hygienic and protective merits and condition, used in the development of measures for the architectural planning of the territory, care of plantations and improvement of the existing landscapes.

The main task of landscape taxation consists in revealing in nature and plotting special landscape divisions with homogeneous architectural, artistic and biological characteristics [10].

Taxation indicators of plantations are the system of taxation features of plantations that determine their quantitative and qualitative assessment, biological and physical features of their structure and performance under certain conditions within the area they occupy. Taxation indicators are the basis for taxation description in the process of green space taxation.

The set of taxation indicators includes: the origin of stands (natural and artificial); form - simple (one-tier) or complex (multi-tier - the crowns of trees form several tiers); composition - the ratio of tree species forming the stand; average height and average diameter of the stand, stand age, quality class, completeness, wood supply, marketability class, ground cover.

Plantations under the influence of negative natural and anthropogenic factors are additionally characterized by indicators of the biological stability of the plantation.

During taxation there are used measuring tools (measuring fork, altimeter, fullness meter), as well as normative materials (standard table of completeness and stocks, bonitet scale, etc.). Field records of all these indicators and data of measuring tools for each taxation point are made in the taxation card.

According to the existing methods of taxation in urban green space trees are determined by breed, condition, measuring the diameter of the trunk at a height of 1.3 m. According to the diameter an approximate age of the tree is established, because in urban conditions it is impossible to cut model trees and use an age drill.

The trunk diameter of individual trees can be measured with a measuring fork or a tape measure. Before starting work, use a compass to determine north. Measurement of trunk thickness with a measuring fork is made in two mutually perpendicular directions north-south and west-east. Accuracy of readout should be up to 0,1 cm, at mass recalculations up to 1 -2 cm. diameter is determined as an arithmetic average of two measurements.

The age of trees is determined by the number of annual layers or by eyeballing the following features:

- on the development of the crown:

    a) The younger generation has a sharper crown;

    b)older spherical crowns;

    c) close to natural ripeness - umbrella-shaped.

- by the whorls and their traces in young pine trees

- the leaves and needles of young trees differ markedly in color and size; the crowns of old trees are thinning

- bark and crust are differently developed, differing in pattern and color

- older trees are cleaner and higher cleared of limbs [11].

Evaluation of shrubs is to count their number by species. A scheme of the taxidermied object with the boundaries of lawns and flower beds is made, and the location of trees is also indicated there.

The majority of cities of the Russian Federation did not carry out a complete inventory of plantations, as the municipal authorities were unable to allocate funds for this purpose. However, these works are carried out in a fragmented manner, mainly on the most valuable objects of green economy [12].

Record keeping of green spaces is carried out on the basis of materials of the inventory of green spaces at landscaping facilities, materials of forest management and other types of survey of areas occupied by green spaces. All types of green spaces: trees, bushes, lawns, flower beds are subject to accounting.

Record-keeping of green spaces in the city of Vologda is carried out within the boundaries of the registration area in order to determine their number, species composition and condition.

An accounting plot is a land plot that has established boundaries and is provided for use, possession, disposition institutions, organizations, enterprises or individuals - responsible owners.

Inventory and other methods of accounting green areas in the accounting area are carried out in accordance with the methodological documents approved in the prescribed manner methods of subsurface accounting, contour recounts, taxation, depending on the complexity of the structural parts of the green areas located within the boundaries of the accounting area.

The document reflecting the results of the inventory of green spaces is a passport of the inventory area.

The passport of the registration area contains the following obligatory information:
- an inventory plan of the territory;
- administrative-territorial affiliation of the registration area;
- the name of the responsible owner;

- the established regime of urban planning activities;
- the established functional purpose of the land plot;
- total site area;
- the number of green spaces;
- species composition of green spaces;
- the condition of green spaces.

Aims and objectives of keeping green space records:
- Obtaining reliable data on the species and age composition, quantitative and qualitative characteristics of green spaces on the territory of Vologda.
- Determining the appropriateness of   the activities   carried out responsible owners on landscaping facilities, the established functional purpose of the areas.
- Analysis of the condition of green spaces in the city.
- Creating an information base for the organization of the rational use of landscaping objects in Vologda.

Inventory is carried out in accordance with the instructions for the inventory of green spaces in cities, working, holiday and resort settlements [13]. The best time for this work is spring or early fall. The initial data for the inventory of the object is the existing master plan of the territory in M 1:500 (1:200) or executive drawings (planting, dividing) on the basis of the geodetic plan.

Work on the inventory of massive urban facilities is usually performed by the Bureau of Technical Inventory (BTI) of the district, city, but with the obligatory involvement of specialists in landscape,   dendrological and   entomophyto-pathological examination of green spaces, identifying the condition of the site as a whole, violations of the planning network, types of spatial structure, types of park plantings.

Work on the initial inventory for the purpose of further development of the project of reconstruction of plantings on the site is funded in full for public facilities - at the expense of the local budget, and for departmental facilities - at the expense of the departments specifically provided for in the estimates for landscaping.

The organizations conducting the contract inventory make and keep original materials on the accounting of green areas for each object, giving customers the required number of copies [15].

Methods of gardening inventory of plantings include:

1) a continuous plant enumeration or a detailed survey and assessment of each woody plant in separate and responsible areas of residential development;

2) landscape-taxation assessment of plantings in areas with a high density of plants, their excessive densification and chaotic placement (mainly at the ends of buildings, from the rear of the facades of buildings, etc.)

In the landscape-taxation method large and valuable large-aged specimens of trees are marked and described in detail; the main vegetation, which is not of special value, is evaluated in generalized terms, "clumps" are shown on the plan as a partition with conventional signs of plant species ("B" - birch, "L" - linden, etc.). Plots with stands are described by species composition, age, completeness, average height, undergrowth, and soil cover.

In the landscape-taxation method of inventory when describing the following provisions are highlighted :

- plantation composition;
- planting age;

- the characteristic of the sanitary condition;
- aesthetic condition (assessment of decorativeness);
- preliminary measures for reconstruction (thinning, removal of low-value specimens, sanitary cleaning, etc.)

In a detailed inventory, each plant is numbered and then mapped on the plan. In situ anchoring of plants is carried out to the existing elements of the layout - to the borders of the driveway or to the paved path, to the backing of the building, to the structure. Binding is carried out with the help of a tape measure, as well as measured steps. The data obtained by natural measurements of plants are recorded on the working plan in M:100.

Continuous counting of stands is carried out only on sites where their number does not exceed 300 units (roadside forest plantations, special purpose forest plantations, intradistrict plantations). In the city of Vologda, continuous recalculation was used in the inventory of stands in the square on Babushkin Square, Boulevard on Victory Avenue, the square on the Kopanka River.

At sites with more than 300 trees, instead of a continuous tree count, linear or rectangular sample areas of 400 square meters are laid, on which a continuous count of woody plants is carried out. After collecting and processing plantation data on the trial area, a characteristic of plantations on the whole site is formed.

Inventory of green areas and structural elements of the object is carried out in two stages:

The first is a field one;

The second is the cameral processing of the subsumed material.

At the first stage, work is carried out on the study of existing documentation, clarification of the boundaries of the object in the red lines and landscape planning data, existing types of spatial structure, the study of the presence of communications and structures and conduct a survey or

additional mapping of plantations from nature to the plan with the corresponding records in the work log.

At the second stage, the received data is summarized, the logs and records are analyzed and put in order, a balance of the territory is developed, an inventory plan is drawn up, a corresponding certificate of completion is drawn up for their acceptance - handover. On the basis of the received materials, the passport of the object is updated (corrected). Depending on the size of the object and availability of green spaces, the inventory can be carried out in different ways.

For large parks and forest parks the work is carried out by a brigade method - by a special taxation team using the method of landscape taxation. For urban landscaping objects - squares, boulevards, gardens, areas of residential development - the work is done in an individual way, by plotting each plant, all types of green spaces and structural elements.

When recounting the trees, their species, diameter, condition category, damage by pests, diseases and other negative (including anthropogenic) environmental factors are indicated. If necessary, notes are made on the peculiarities of the crown structure of woody plants and the methods of shaping or deep pruning applied to them.

The following parameters and indicators of trees are indicated in the recalculation list (work diary):

- a type of woody plant;

- trunk diameter (cm) at a height of 1.3 m;
- tree height (m);
- tree age group (age class 1-5): under 15 years, 15-25 years, 25-45 years, 45-60 years, and over 60 years.

The category of tree condition is determined according to Table No. 1.

Conifers

Tree condition categories

Category trees	The main signs	More signs
1.No sign of weakening	The needles are green, shiny, the crown is dense, the growth of the current year is normal for this breed, age, conditions. habitat and season.	
2. weakened	The needles are often lighter than usual, the crown is weakly auricular, the growth is less not more than ^ compared to the norm.	There may be signs of local damage trunks and Root Paws.
3. Severely impaired	The needles are light green or light, matte. Crown openwork, growth reduced by more than ^ compared to normal.	Attempts to settle of pests in the arboretum and branches. Possible signs damage and Root Paws.
4. Shrinking	The needles are gray, yellowish or yellow-green, the crown is noticeably thinned, the growth of the current year is still visible or missing.	Attempts to settle pests on trunks and branches. Possible signs of local injury trunk and root feet, only stronger than in the

		of the previous category.
5. The current year's deadwood	The needles are gray, yellowish or brown, the crown is often thinned, small branches are preserved, and the bark is preserved or partially crumbled.	Attempts to settle pests on trunks and branches. Possible signs damage to the trunk and root paws. Insect holes on the trunk.
6.Dry years of previous years	The pine needles had fallen off or remained only partially intact. Small and most of the large branches have fallen off.	On the branches and on the trunk there are holes, under the bark there is abundant brown flour or fungus wood-destroying fungi.
Hardwoods		
0. No sign of weakening	The foliage is green, shiny, the crown is dense, the growth of the current year is normal for this breed, age, habitat conditions and season.	
Weakened (up to 25% of dry branches in the crown)	The foliage is green , the crown weakly, incremental is attenuated compared to normal.	Local damage to branches, root feet and trunk.
2. weakened (в crown of 25% to 50% dry. branches)	Leaves are smaller or lighter than usual, prematurely falls off, the crown is thinned.	Signs previous categories expressed stronger, attempts settlements pests.

3. Severely weakened (50% to 75% of dry branches in the crown)	Leaves are smaller or lighter than usual, prematurely falls off, the crown is thinned.	Signs previous categories expressed stronger, attempts to settlements pests, noted sap flow.
4. Shrinking Dry branches (>75% of dry branches)	Leaves are smaller, lighter or more yellow, prematurely The crown is severely thinned, and it falls off or wilts.	The trunk and branches show signs of pest infestation . Juice bumps.
5. Drywood for the current year	The leaves have withered, wilted, or fallen off prematurely, but the small twigs and bark are still intact.	The trunk, branches and root paws show signs of infestation by stem pests and fungal infestation.
6. Drywood past years	Leaves and part of the branches have fallen off, bark is destroyed or fallen off on most of the branches.	On the branches and on the trunk there are holes of insects, on the bark and under the bark the fungus and fruiting bodies of fungi.

According to the recalculation list (work diary), a summary table of the inventory of green spaces is formed, which allows you to judge the stability of a given plantation.

The characteristic of the object of the inventory of public green spaces is accompanied by the filling of an accounting card (passport), which specifies the necessary data. In the record card (passport)

The following data should be specified : the object code, namely:

01-forest parks, wooded cottages, areas of botanical gardens;

02-parks, landscaped areas of health and sports complexes, recreational areas;

03-Inner yard plantings;

04-boulevards, squares, etc;

05-Simple street plantings;

06 Plantations of sanitary protection zones (including bank protection forests).

And also the area of the city where the object is located, the name of the object, the generalized information on the account trees.

All spatial data and factual information obtained during the inventory of green spaces is integrated into the unified information-analytical complex of the city (electronic version). In order to ensure integration, all data obtained during the inventory of public green spaces must comply with the requirements for the information resources of the information-analytical complex.

The inventory plan should show:

- the outer boundaries of the object;

- the external situation beyond the borders;

- the boundaries and numbers of the registration sites and biogroups;

- location of small architectural forms (schematic);

- placement of lawns, flower beds;

- planar structures and road and path network, taking into account the types of coatings;
-conditional designations and explication.

Particularly valuable tree species (unique, historical) are marked on the plan and numbered in red by separate numbers throughout the object.

On the inventory plan of landscaped and landscaped areas of streets,   passages, alleys, squares, embankments
each tree and its number is shown.

On the inventory plan of the park with a low recreational load or where the flow of visitors is strictly regulated by architectural and planning techniques (historical,   forest parks,
The following situations are shown: glades, glades, reservoirs, etc.). Tree and shrub vegetation   is shown in conditional

notations. In this case, the inventory plan can be made at a scale of 1:1000 and smaller.

In squares, gardens, boulevards, parks with high recreational loads, intradoor and neighborhood plantings, all trees, shrubs (alley plantings), hedges, flower beds and lawns, clumps of group planting of trees and shrubs are drawn on the plan.

In the process of surveying the green spaces located in the survey area, the following data is recorded in the work diary with respect to:

trees located in the passageways:

- type of planting (row planting, group planting);
- tree numbers;
- breed;
- age;
- diameter;
- trees that are pruned are marked;
- condition.

J trees located in squares, gardens and boulevards are recorded the same data as in the passages, except the numbering;

J trees located in the record plots of parks and forest parks:

- type of plantation;
- the predominant composition of the rocks;
- closeness of plantations;
- the number of trees per 1 hectare of area;
- middle age;
- condition.

shrubs:

- type of planting (alley planting, group planting);
- breed;
- age;

- number of bushes;

- length for row (alley) planting;

- condition.

Lawns and flower beds are counted by area (perennial flowers, in addition, are counted by the number of bushes in the accounting area).

The condition of the plantation is determined by the following attributes:

1 - *"good"* - plants are healthy with a regular, well-developed crown, without significant damage; lawns without lacerations and with well-developed grass - mowed or meadow, flower beds without wilted plants and their parts (Fig. 3.1);

2 - *"satisfactory"* - plants are healthy, but with improperly developed crowns, with significant, but not life-threatening wounds or damage, with hollows, etc.; shrubs without weeds, but with the presence of shoots; lawn with small weeds, poorly maintained herbage; flower beds with the presence of wilted parts of plants (Fig. 3.2);

3 - *"unsatisfactory"* - stands with irregular and poorly developed crowns, with significant damage and wounds, with infestation of diseases or pests that threaten their life; shrubs with the presence of shoots and dead parts, with weeds; lawns with rare, dying out, full of weeds grass stand; flower beds with large fallout of flowers, wilted plants and their parts (Fig.3.3).

Work on the inventory of massive urban facilities is usually performed by the Bureau of Technical Inventory (BTI) of the district, city, but with the obligatory involvement of specialists in landscape, dendrological and entomophyto-pathological examination of green spaces, identifying the condition of the site as a whole, violations of the planning network, types of spatial structure, types of park plantings.

Work on the initial inventory for the purpose of further development of the project of reconstruction of plantings on the site is funded in full for public facilities - at the expense of the local budget, and for departmental facilities - at the expense of the departments specifically provided in the estimates for landscaping.

The organizations conducting the inventory under the contracts make and keep original materials on the inventory of green spaces for each object, giving customers the required number of copies of them [15].

The area of the inventoried object is calculated from the site plan using one of the following methods:

1) by breaking it down into the simplest geometric shapes;

2) planimeter;

3) palette (small area contours);

4) analytically [13].

The best time to conduct a green space inventory is spring or early fall.

Objects are inspected once every five years in order to identify changes in the internal situation and reflect them in the inventory materials: on the inventory plan and in the passport of the object [14].

Thus, according to the above methodology, an inventory of green spaces in the square on the Kopanka River was carried out.

The square was laid from 1960 to 1980 between Zosimovskaya Street and Galkinskaya Street. July 20, 1984 in the park opened a monument-monument of 85 millimeter cannon of the Great Patriotic War.

During the inventory of the park area the following work was done: on the site plan at a scale of 1:500 point objects corresponding to the location of plants in the landscaped area of the park were marked. Each tree was given an individual number; information about the plant was entered into a pre-made inventory card.

The species name and serial number were noted; the circumference of a tree at the height of 1.3 m was measured and the diameter was calculated; the height of tree and shrub vegetation was measured with an eclimeter; the drying of bushes and crowns was determined visually in %; the largest mechanical damages of plants were noted, the number of untreated cuttings was specified. Infected plants, mechanical damages were noted, and shrub trimming was indicated.

Based on the information obtained, a technical passport for the square was prepared.

According to the results of the inventory calculated that the area of the square is 13033.18 square meters, including under the trees 8610.80 square meters under the asphalt surface 4110.38 square meters, under the paving 312.00 square meters.

There are 12 species of trees and three types of shrubs in the garden. The dominant species is the birch. The average age of the plant is 35-40 years, the wear is up to 55-60%.

Shrubs are trimmed in order to preserve their ornamental value, increase the number of flowers (fruits), enhance growth, improve the health of plants, as well as to regulate their size. It is possible to arrange competitions on decorative decoration of trees along public gardens or along the roadway, which will bring a lot of variety to the city.

Every year, the city's flower gardens should be developed and expanded. Flowerbeds are very decorative for parks and streets, it is quite a simple way to decorate the city. Tiered flowerbeds or flowerbeds with a specially designed design look beautiful. Small flowerbeds near residences and organizations are possible. Flowerbeds can be a variety of shapes and consist of a different set of flowers, differing not only in plant height and color of the flower, but also different periods of flowering.

Lawns are the first step in landscaping planning. However, the age of the lawn is a decisive factor in its quality. Unfortunately, urban areas, unlike private estates, cannot expect to pay close attention to lawn care. Any lawn, if limited to lawn care by seeding and occasional mowing, will overgrow with weeds in the early years when the grass plants of the lawn compete for nutrients and space with the weeds. As a result, you may see lawns of poor quality throughout the city. Not all citizens understand the meaning of timely mowing, allowing its absence increases the competition between the plants of the lawn, which clearly leads to the loss of plants and the formation of mounds. For lawn care, you need professionals who will be engaged in lawn improvement. This, of course, requires additional funds allocated from the regional budget.

The main focus of work remains the overhaul and improvement of public landscaping.

A comprehensive landscaping plan shall contain the following specific indicators and measures:
- plan for the participation of botanical gardens and arboretums-reserves in the procurement of material to expand and enrich the range of planting material;
- the training plan for urban greening specialists;

- mechanization of green building and organization of mechanized leskhozes;
- mechanization of the care of green spaces;
- sanitary and anti-malaria measures;

- plan for breeding birds and animals in forest parks and forests and fish in water bodies adjacent to the city limits;

- fire prevention measures;

- control of pests and diseases of green spaces;

- protection of green areas and water bodies;

- rules for the use of suburban forests, forest parks, parks, beaches, reservoirs and other places of mass recreation among green spaces;

- plan for scientific and research work on forest park construction, landscape architecture, landscape gardening, floral design, floriculture, etc;

- plan for the design of typical structures of structural elements of integrated urban greening systems.

Conclusion

During the analysis of green spaces in Vologda it was found that the green spaces are distributed unevenly across the planning structures of the city. The public greenery is poorly represented in the Zarechny part. The largest area of green space is in the central district.

The species composition of ancient Vologda gardens, boulevards, and neighborhood plantings is very poor - about 20 species of trees and shrubs: various poplars, lilac, hawthorn, elm, lime, birch, yellow acacia, occasionally larch, fir, ash, and oak. Although the natural conditions of Vologda allow the use of a much larger number of tree and shrub species,
that would ensure the proper architectural and dendrological
city design

The inventory is subject to all green spaces, regardless of the organizational and legal form of ownership and departmental affiliation, located within the city limits, with established boundaries and provided for use (possession, disposal)   to responsible land users   (institutions,
organizations, enterprises or individuals).

List of references

1. Shver, C.A. Climate of Vologda / ed. by C.A. Shver, A.S. Egorova.-L.: Gidrometeoizdat, 1988.-174 p;

2. [Electronic resource] . - Access mode: http://history.nason.ru/klimat/;

3. [Electronic resource] . - Access mode: http://history.nason.ru/priroda/;

4. Suslova T.A., Repina N.N. Green attire of Vologda / T.A. Suslova, N.N. Repina // Vologda: a local history almanac. - Vologda, 1997. - Vol. II. - C.764.;

5. Erokhina V.I., Zherebtsova G.P., T.I. Wolftrub. - M.: Rosselkhoizdat, 1987.- 480 p;

6. Gorokhov, V.A.   The green nature of the city: a textbook for universities / V.A. Gorokhov.-M.: Architecture-S, 2005.-528s;

7. Construction Norms and Rules of the Russian Federation. Urban Planning. Planning and Building of Urban and Rural Settlements: SNiP 2.07.01-89*;

8. Russian Federation. Laws. Urban Planning Code of the Russian Federation: Federal Law dated 29.12.2004 N 190-FZ;

9. On General Principles of Organization of Local Self-Government in the Russian Federation: Federal Law No. 131-FZ of October 6, 2003;

10.   Nikolaevskaya, I.A. Landscape taxation and the formation of plantations of suburban areas / I.A. Nikolaevskaya. - L.: Stroyizdat, 1977.- 224 p.;

11. Fundamentals of landscape gardening and landscaping settlements: guidelines and assignments for laboratory work, part 1. - Vologda: Vologda State Technical University, 2007;

12. Luntz, L.B. Urban green building: a textbook for universities / L.V. Luntz.-M.: Stroyizdat, 1974. - 275 p;

13. Russian Federation. Laws. Instruction on inventory of green spaces in cities, workers, dacha and resort settlements of the RSFSR: approved by Order of the Ministry of Agriculture of the RSFSR March 12, 1971 N 130;

14. Teodoronsky, V.S. Construction and operation of objects of landscape architecture: textbook for students of higher educational institutions / Teodoronsky, V.S., Sabo E.D., Frolova VA - M. : Publishing Center "Academy", 2006.- 352 p;

15. [Electronic resource] . - Access mode: http://stroy-sad.ru/organizatsiya-rabot-v-sadovo-parkovom-

stroiteistve/ inventarizatsiya-na-sadovo-parkovyih-obektah/2/

16. [Electronic resource] : http://www.sdelaemsami.ru/index.html;

17. Udilov, VN Handbook of engineer lespromkhoz on labor protection and safety / VN Udilov.-Moscow: Forest Industry, 1975.- 230 p;

18. V. Popov, Y.V. Ensuring safety in the fight against pests and diseases of the forest / V.V. Popov. - M.: Forest Industry, 1978.- 27 p..

Printed by Books on Demand GmbH, Norderstedt / Germany